Curious Creatures

TARSIERS

GAIL TERP

BLACK RABBIT BOOKS

Bolt is published by Black Rabbit Books
P.O. Box 227, Mankato, Minnesota, 56002
www.blackrabbitbooks.com
Copyright © 2023 Black Rabbit Books

Marysa Storm, editor; Michael Sellner, designer and
photo researcher

Library of Congress Cataloging-in-Publication Data
Names: Terp, Gail, 1951- author.
Title: Tarsiers / by Gail Terp.
Description: Mankato, Minnesota : Black Rabbit Books, [2023] |
Series: Bolt. Curious creatures | Includes bibliographical references and index. |
Audience: Ages 8-12 | Audience: Grades 4-6 | Summary: "Tarsiers are animals
unlike any other, and curious and reluctant readers alike will be captivated by
these exceptional creatures' life cycle, habitats, diet, and threats to survival
through carefully leveled text, detailed infographics that emphasize visual literacy,
and engaging, colorful photography"– Provided by publisher.
Identifiers: LCCN 2020017985 (print) | LCCN 2020017986 (ebook) |
ISBN 9781623105693 (hardcover) | ISBN 9781644664940 (paperback) |
ISBN 9781623105754 (ebook)
Subjects: LCSH: Tarsiers–Juvenile literature.
Classification: LCC QL737.P965 T47 2023 (print) | LCC QL737.P965 (ebook)
| DDC 599.8/3–dc23
LC record available at https://lccn.loc.gov/2020017985
LC ebook record available at https://lccn.loc.gov/2020017986

Image Credits

Contents

CHAPTER 1

A Tarsier in Action

It's just after sundown. A tarsier clings to a tree. Its huge eyes watch for **prey**. Its big ears listen. The little **primate** hears something. It's an insect on a nearby branch. The tarsier pushes off the tree with its long back legs. It lands on the branch and snatches up the bug. Time to eat!

Tarsier is pronounced TAHR-see-er.

COMPARING WEIGHTS

6
5
4
3
2
1
0

1.7 to 2 OUNCES
(48 to 57 grams)

Giant Eyes

There are about 10 kinds of these small **predators**. They all have huge eyes. In fact, tarsier eyes are so big, the primates can't move them. That's OK, though. They can turn their heads far enough to see behind them. Their large ears can also turn toward sounds.

Philippine tarsier	spectral tarsier	western tarsier

2.8 to 5.8 OUNCES
(79 to 164 g)

3.6 to 4.6 OUNCES
(102 to 130 g)

3.7 to 4.7 OUNCES
(105 to 133 g)

PARTS OF A TARSIER

HUGE EYES

MOUTH

EARS

NOSE

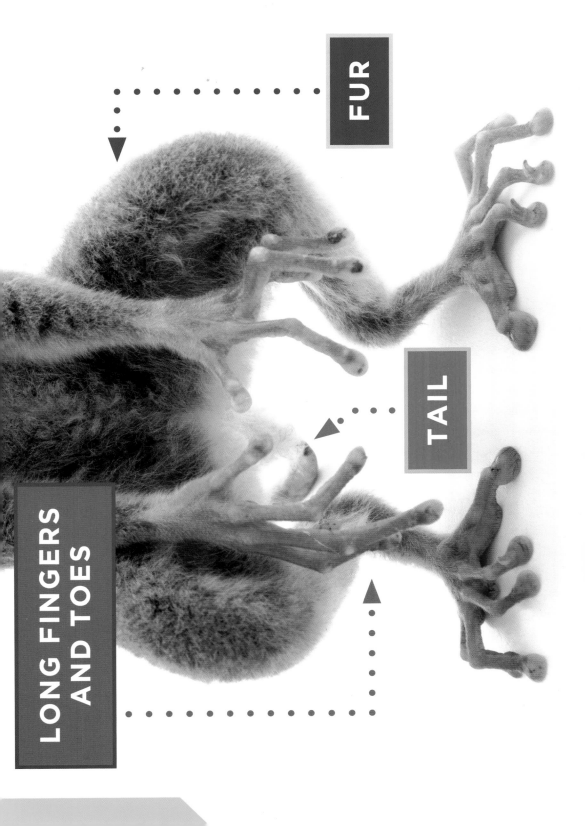

FUR

TAIL

LONG FINGERS AND TOES

CHAPTER 2

FOOD
and Home

Tarsiers are **nocturnal**. Throughout the night, they travel and hunt. Tarsiers hunt mostly in trees, grabbing bugs from branches and leaves. Sometimes they search for prey on the ground. These animals often eat insects. They also eat birds, bats, and small lizards and snakes.

Tarsiers are the only primates that do not eat plants.

Tarsier Habitats

SCRUB FORESTS

MOUNTAIN FORESTS

FARMS • • • •

Where Tarsiers Live

Tarsiers are found only on islands in Southeast Asia. They live in places with many trees. Tarsiers need the trees to hunt from and sleep in. Tarsiers are shy. Most of them live far from humans. But not all. Some will live near farms.

North
America

South
America

Europe

Asia

Africa

Australia

CHAPTER 3

The Family

Not all tarsiers have the same kinds of families. Some types of tarsiers live alone when they're adults. Others stay in pairs or family groups. Tarsiers in groups play and groom each other. To **communicate**, they use chirps, whistles, and other sounds.

Tarsiers send and hear **ultrasonic** sounds. They make high-pitched sounds to warn each other of danger. And they listen for the sounds their food makes. Most other animals can't hear these sounds.

COMPARING WEIGHTS

NEWBORN PHILIPPINE TARSIER

ABOUT 1 OUNCE (28 g)

ADULT PHILIPPINE TARSIER

ABOUT 4 OUNCES (113 g)

Babies

Some types of tarsiers **mate** in spring and fall. Others have no special mating season. About six months after mating, females give birth to one baby. Babies are born with fur. Their eyes are open. Most babies can climb on their first day. Mothers still carry their young for the first few weeks, though. Some carry their babies in their mouths like cats do.

Growing Up

Babies begin to hunt in about a month. But their mothers still let them **nurse**. In about two months, the young stop nursing. They can hunt for themselves. Types that live alone leave their mothers. Others might stay with their groups.

By the NuMBers

0.6 INCH
(15 millimeters)

width of a tarsier eye

34

number of teeth

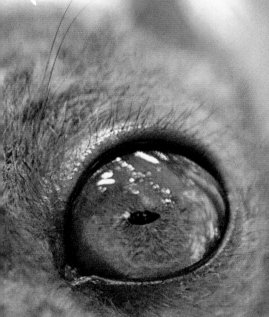

about
16 FEET
(5 METERS)

**HOW FAR
SOME TARSIERS
CAN LEAP**

10 YEARS
LIFE SPAN

180
DEGREES how far a tarsier can turn its head in one direction

CHAPTER 4

Danger!

Tarsiers have many predators. Some, such as owls, come from the air. When groups spot attackers in the sky, they call an alarm. The tarsiers then scatter and hide. Other predators are in trees or on the ground. Tarsiers must watch out for snakes and monitor lizards. Groups of tarsiers will **mob** snakes and lizards. They make noises and bite. Tarsiers in groups don't make easy meals.

Tarsier Food Chain

This food chain shows what eats tarsiers.
It also shows what tarsiers eat.

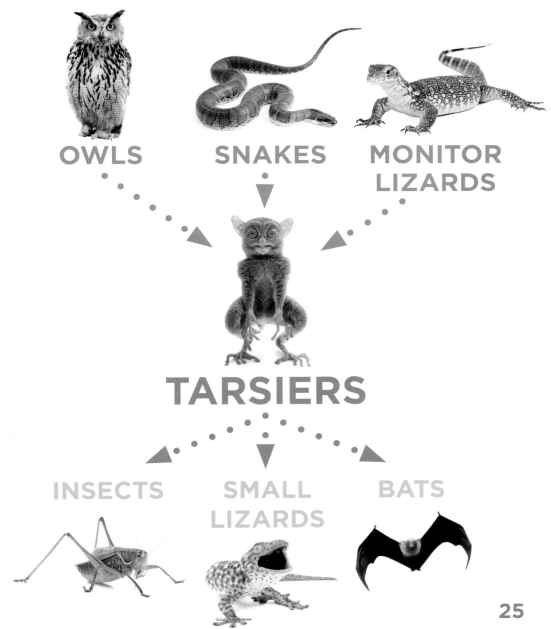

OWLS　　SNAKES　　MONITOR LIZARDS

TARSIERS

INSECTS　　SMALL LIZARDS　　BATS

TRAPPED

DEFORESTATION

POLLUTION

Tarsiers and Humans

Humans are also a threat to tarsiers. They hunt tarsiers for food. They trap the little animals and sell them as pets. Humans also clear trees for logging and farming. Cutting down trees leaves less land for the animals to live on. Pollution is a problem too. It causes tarsiers to get sick and die. Tarsiers are now in danger of dying out.

Protecting Primates

Some people work to protect these creatures. The Philippine Tarsier Foundation has set aside land where tarsiers can safely live. No hunting is allowed there. These spots also give scientists a place to study tarsiers. By studying tarsiers, people hope to find new ways to protect them. Tarsiers play a big part in their habitats. People need to help keep these curious primates from disappearing.

People have made laws against keeping them as pets.

GLOSSARY

communicate (kuh-MYU-nuh-kayt)—to share information, thoughts, or feelings so they are understood

mate (MAYT)—to join together to produce young

mob (MOB)—to crowd about and attack or annoy

nocturnal (NOK-turn-uhl)—active at night

nurse (NURS)—to drink milk from the mother's body

predator (PRED-uh-tuhr)—an animal that eats other animals

prey (PRAY)—an animal hunted or killed for food

primate (PRI-mayt)—any member of the group of animals that includes humans, apes, and monkeys

ultrasonic (uhl-truh-SON-ik)—used to describe sounds that are too high for humans to hear

BOOKS

Beer, Julie. *Bite, Sting, Kill.* Washington, DC: National Geographic Kids, 2023.

London, Martha. *Asia.* World Studies. Lake Elmo, MN: Focus Readers, 2021.

Shumaker, Debra Kempf. *Peculiar Primates: Fun Facts about These Curious Creatures.* Philadelphia: Running Press Kids, 2022.

WEBSITES

Philippine Tarsier
animalia.bio/Philippine-tarsier

Tarsier Facts for Kids
kids.kiddle.co/Tarsier

The Cutest Little Predator
www.youtube.com/watch?v=_oTK9NsrmvM

INDEX